YOUR KNOWLEDGE HAS VALUE

- We will publish your bachelor's and
  master's thesis, essays and papers

- Your own eBook and book -
  sold worldwide in all relevant shops

- Earn money with each sale

Upload your text at www.GRIN.com
and publish for free

Bibliographic information published by the German National Library:

The German National Library lists this publication in the National Bibliography; detailed bibliographic data are available on the Internet at http://dnb.dnb.de .

Imprint:

Copyright © 2018 GRIN Verlag
Print and binding: Books on Demand GmbH, Norderstedt Germany
ISBN: 9783668894563

This book at GRIN:

https://www.grin.com/document/457608

Leon Cena

On the Success of the Economic Area Frankfurt-Rhein-Main

Economic Structure and Location Factor

GRIN Verlag

Research paper

THE ECONOMIC AREA FRANKFURT-RHEIN-MAIN – WHY IT IS SUCCESSFUL AND IS THE POTENTIAL EXPLOITED?

Leon Cena

Table of contents

1. Introduction

When I thought about the topic of my research paper, I quickly decided I want to write about economy because I am interested in that topic. During my research, I came across the city, Frankfurt am Main. The facts that Frankfurt has a good working economy and an impressive skyline, which is unusual for German cities compared to American cities, were known to me. I asked myself why Frankfurt was so successful and if it may still develop, but I could not answer these questions. The main topic of this research paper is to answer this question.

At the beginning of my research paper, I would like to describe the location of the economic area of Frankfurt-Rhein-Main and consider which cities are members of this economic area. I will explain important location factors of the tertiary sector and examine Frankfurt am Main regarding the present location factors. I will do that with the help of the book "Wirtschaftsgeographie", two editions of the book "Wirtschaftsstandort Frankfurt am Main" and several statistic reports.

I will focus on Frankfurt because it is the most significant city in the economic area. Afterwards, I will analyse the economic structure of Frankfurt and Frankfurt-Rhein-Main and compare them with the structure in the past. While analysing the current economic structure, I will elaborate significant branches in the economy of Frankfurt. At the end of this work, I will try to realize a trend and give a prognosis. I will conclude by answering the central issue, why the economic area could be so successful and if the potential may be exploited.

2. Description of the location and administrative structure

The region Frankfurt-Rhein-Main is a German metropolitan region. There are eleven metropolitan regions in Germany. The region was declared by the ministerial conference for Regional planning on April 28, 2005[1].

This chapter contains the localization of the Frankfurt-Rhein-Main region. The region is located between 7,6° East and 9,72° East and 49,4° North and 50,81° North in West-Central Germany and on the continent of Europe. The region extends across great parts of the federal state Hesse and smaller parts in Bavaria and

---

[1] Extracted: https://www.region-frankfurt.de/Regionalverband/Region-in-Zahlen/Statistik-Viewer-Regionalverband/index.php?mNavID=2629.36&sNavID=2629.36&La=1#a2, status as of February 13, 2018

Rhineland-Palatine[2]. The Rhein-Neckar metropolitan region is a neighbouring metropolitan region. Neighbouring federal states are Thuringia in the East and Stuttgart in the South.

The region consists of seven independent cities and 18 districts. For a detailed table of the cities and districts, see figure 2. The North-South expansion is 155 kilometres, and the East-West expansion is 165 kilometres[3]. The total area of the region is 1.475.530 square metres[4].

3. Important location factors of the tertiary sector

This chapter concentrates on important factors of the tertiary sector. Every company must choose a certain location for their business operation. Entrepreneurs responsible for the company start-up must choose a location for their first office or production site. Other companies may think about additional offices, production facilities, or a resettlement. They all need suitable locations. This raises the question why companies search for a certain location and how they can determine the correct location.

The companies want to settle down at the location that fulfils the requirements they have. The aim is to find the location with the best possible conditions. Every company has specific requirements that should be met by the location. To find a suitable location, the companies must think about the location factors. The location factors must be compatible with the requirements of the company[5]. While figuring this out, the companies can filter out unsuitable locations. This decision is considered important because a resettlement can be expensive. As an effect, these decisions are often relatively firm.

The location requirements of a company depend on the sector and branch. A software company does not have the same requirements as a farm, for example.

Before explaining the important factors of the tertiary sector, it is important to differentiate the location factors. There are hard and soft location factors. Soft location

---

[2] Cf. figure 1
[3] Measured with the online service Google Maps
[4] Cf. Stadt Frankfurt am Main, Bürgeramt, Statistik und Wahlen. Strukturdatenatlas Frankfurt, Frankfurt am Main, 2015
[5] Cf. Palme, G. & Musil, R.: Wirtschaftsgeographie, Braunschweig 2012, page 11

factors cannot be easily calculated, in contrast to the hard location factors, which can be precisely calculated[6].

An important location is the supply of labour. Whereas the price of the labour is important in the first and secondary sector, the quality of the labour is important for the tertiary sectors. This is especially the case for high-tech companies or other high valued services, like financial services or consultancy. The companies are willing to pay high salaries and expect good work (see chapter 5.2). These highly skilled workers make demands on the employer. They demand good living standards and environmental conditions[7]. Because of that, the soft location factors are often very important for the tertiary sector. Companies must attract the most qualified workers, and they can do that only by offering good conditions. Other soft location factors are culture, good education, and availability of leisure activities.

Nevertheless, soft location factors are especially important for the tertiary sector, and the hard location factors are important as well. Important hard location factors are the availability of land, labour, traffic infrastructure, communication infrastructure, proximity to the costumer, and financial incentives. Area is important for companies that offer services. Banks and law firms, for example, need many offices because they have many workers. Production facilities are not as important for the tertiary sector as they are for the secondary sector. Agglomeration is also advantageous for service companies. When there is a high concentration of companies of the same branch, cooperation is possible. The proximity of research facilities and universities is also very important, especially for high-tech companies, because these companies have a high demand for research and qualified workers.

4. Location factors of Frankfurt am Main

This chapter deals with the location factors of the city Frankfurt am Main. The analysis of the factors focuses on Frankfurt because it is the most significant location of the economic area Frankfurt-Rhein-Main, even though there are many other cities and districts contributing to the economic performance of the area. It is crucial to identify and to examine the location factor for answering the first part of the main question. Why is the economic region successful? Successful means, in our terms,

---

[6] Cf. Palme & Musil, 2012, page 15-16
[7] Cf. Palme & Musil, 2012, page 14

the gross domestic product of that specific area is high. To have high gross domestic products, there have to be local companies in Frankfurt that create value. Companies must be attracted to settle down in the city. As explained in chapter three, every company chooses the location with the best possible conditions. So, what makes Frankfurt so successful?

The city is in the middle of Europe. The infrastructure of Frankfurt is excellent[8]. The airport in Frankfurt am Main is one reason the infrastructure is so good. It is the eighth largest airport worldwide[9]. 60.792.308[10] passengers were transported in 2016. The total air freight of 2016 was 2.067.257 tons[10], and there were 468.900 aircraft movements[11]. Almost all the European capitals can be reached within two hours from the airport in Frankfurt[12]. The road and railway network are developed. The total length of the roads in Frankfurt, which consists of federal motorways, federal roads, states roads, and country roads, is 368 kilometres long. The federal motorways A3, A5, A66 and A661 are in Frankfurt[13]. Up to 450.000 travellers use the railway system every day. It is "one of the most important traffic hubs in Europe."[14] For more figures about the railway system, please see figure 3. The public transport system works well with when travelling within the city centre and helps 300.000 commuters to reach their destinations[15] every day. 13 ICE's are connecting other metropolitans with Frankfurt[15]. A harbour port is located in Frankfurt as well. This harbour port is located near the motorways A661 and A661 and the railway system. It is also connected with other important waterways, like the river Rhine in the South-West near Mainz[16]. It is recognizable that the infrastructure is well developed. This is a major location factor of Frankfurt because many people depend on it. 444.920 commuters depend on a good working infrastructure and the other

---

[8] Cf. Kirk, C.: Wirtschaftsstandort Frankfurt am Main, Darmstadt 2001, page 9

[9] Cf: http://www.frankfurt.de/sixcms/detail.php?id=2556771, status as of February 18, 2018

[10] Cf. Stadt Frankfurt am Main, Bürgeramt, Statistik und Wahlen. Statistisches Portrait / Statistical portrait Frankfurt am Main 2016, Frankfurt am Main, 2017, page 6

[11] Cf. Industrie- und Handelskammer, IHK-BEZIRK FRANKFURT IN ZAHLEN 2016|2017, Frankfurt am Main, 2017, page 6

[12] Cf: http://frankfurt-business.net/standort-frankfurt/infrastruktur/, status as of February 18, 2018

[13] Cf: https://www.google.de/maps/place/Frankfurt+am+Main/@50.1213152,8.3563599,10z/data=!3m1!4b1!4m5!3m4!1s0x47bd096f477096c5:0x422435029b0c600!8m2!3d50.1109221!4d8.6821267?hl=de, status as of February 18, 2018

[14] Extracted: http://www.frm-united.de/en/location-advantages/frankfurtrheinmain-as-a-business-location/international-traffic-hub/, status as of February 28, 2018

[15] Cf: https://exporo.de/standortanalyse/frankfurt-am-main, status as of February 18, 2018

[16] Cf: https://www.google.de/maps/search/Frankfurt+am+Main+Hafen/@50.0981117,8.5625057,12z , status as of February 18, 2018

people in Frankfurt that use the public transport system. The digital infrastructure is extremely well-developed. The second most important and largest internet access point and hub is in Frankfurt[17]. The availability of glass fibre for private persons and enterprises is extensive[18]. This is important for the big companies in Frankfurt. These companies often have large office buildings and need a high bandwidth. This is technically only possible with glass fibre and is especially important for companies in the IT branch because they have a higher bandwidth for their service in contrast to other branches. This is a location requirement for most IT companies and given to a great extent in Frankfurt am Main.

This infrastructure "guarantees a rapid access to all-important key markets in Europe"[19] and makes the location attractive for customers and employees.

The reputation of Frankfurt is very good. Frankfurt is known for their world-class finance centre, service, and exhibition centre. The gross value added of the economic area Frankfurt-Rhein-Main is 8% of the one of Germany[20]. This shows the significance of the economic performance of this region and supports the good reputation of the region. The gross domestic product per capita in Frankfurt is 98.500€[21]. As a comparison, the gross domestic product per capita of Germany is 42.161,3€[22].

This is a good incentive for potential employees. This shows the high productivity of Frankfurt and justifies its good reputation. The reputation is an important soft location factor, even though it cannot be measured. Many workers and companies would prefer a city with a good image over one with a bad image.

Another important location factor is education. Ten[23] universities are located in Frankfurt. The Goethe-University, which is highly renowned, is a good example. 28[21] universities are located in Frankfurt-Rhein-Main. There are 58.000[21] university students in Frankfurt and 192.500[21] university students in Frankfurt-Rhein-Main. These figures show the great education opportunities. This is important for an

---

[17] Cf: http://global-internet-map-2012.telegeography.com, status as of February 18, 2018 and figure 4

[18] Private research: information provided by a local Telekom store

[19] Cf. Kirk, 2001, page 9

[20] Cf. IHK-Forum Rhein-Main, FrankfurtRheinMain in Zahlen 2018, Hanau, 2018, page 2

[21] Cf. Stadt Frankfurt am Main, Bürgeramt, Statistik und Wahlen. Statistisches Portrait / Statistical portrait Frankfurt am Main 2016, Frankfurt am Main, 2017, page 3

[22] Cf. https://data.worldbank.org/indicator/NY.GDP.PCAP.CD?end=2016&locations=DE&start=1970&view=chart, status as of February 20, 2018

[23] Cf: http://frankfurt-business.net/standort-frankfurt/zahlen-daten-fakten/bildung/, status as of February 18, 2018

economic area dominated by the tertiary sector because the highly demanded workers must be highly skilled and qualified. These diverse education possibilities ensure there are as many potential qualified workers as possible in the region. There are also 13 international schools. This shows the extent of the internationalism of Frankfurt. The companies want to attract the best workers of the whole world. These international schools increase the attractiveness of Frankfurt in the eyes of foreigners because these schools can educate their children, even though they might not speak German fluently. These schools are also important because there are many foreign people in Frankfurt. "Almost 30 % of people in Frankfurt speak a foreign mother tongue."[24]

Research facilities are also important for the tertiary sector. This requirement is fulfilled in Frankfurt. There are 35 research facilities[21].

As mentioned, the labour force is an important location factor. The quality of the labour is important in the tertiary sector. The size of the labour force was 511.633 in 2013[25] and 679.000 in 2015[26]. Clear development underlines the fact that the economy in Frankfurt is successful and still growing. These statistics are referring to the labour force of the year 2013. 93.018[24] workers, about 18,18% of the total workers, are classified as experts. These workers completed at least four years of higher education or have excellent work experience. 95.100[24] or about 18,6% of the workers are classified as specialists. Specialists are foremen who completed an apprenticeship[24]. But, although there is a great labour force, there are still 9.433 notified vacancies[27]. This shows the demand of the labour force is higher than the supply. The skill shortages can be problematic for the companies because they cannot find qualified workers. The data from the "Fachkräftemonitor"[28] of Hesse prove the skill shortages in Frankfurt. It is visible in the figures 5-7 that the demand for economic science, management, and computer science is higher than the supply. Especially, the lack of economic science and management is large. These skill shortages

---

[24] Kirk, 2001, page 11
[25] Cf. Stadt Frankfurt am Main, Bürgeramt, Statistik und Wahlen. Klassifikation der Berufe 2010: Erste Ergebnisse zu den Beschäftigten in Frankfurt am Main am 30. Juni 2013 (Teil 2), Frankfurt am Main, 2014, page 1-2
[26] Cf. Stadt Frankfurt am Main, Bürgeramt, Statistik und Wahlen. Statistisches Portrait / Statistical portrait Frankfurt am Main 2016, Frankfurt am Main, 2017, page 3
[27] Cf. Industrie- und Handelskammer, IHK-BEZIRK FRANKFURT IN ZAHLEN 2016|2017, Frankfurt am Main, 2017, page 3
[28] Cf: http://www.fachkraefte-hessen.de/fachkraeftemonitor.html#ivmhirej2vik, an online tool provided by the Chamber of Commerce and Industry, status as of February 18, 2018

result in higher salaries according to the principle of supply and demand. The importance and popularity of those jobs are underlined by the total amount of graduates in these subjects. There are many graduates in Darmstadt because the Technische Universität Darmstadt, a well-known university in the fields of science, is located there[29]. Considering the most demanded branch was the finance service, it is reasonable and good for the future of the region that there are many universities offering studies in that topic. Most universities with those studies are located in Frankfurt[28].

There are also agglomeration benefits in Frankfurt. As mentioned, there are several universities, research facilities, and libraries. These are all closely concentrated in Frankfurt. Clusters of finance services, consulting services, IT services, telecommunication services, and logistic services are located in mutual proximity. For example, there are more than 300[30] banks and the four largest financial institutions are in Frankfurt. These four financial institutions are the Deutsche Bank, Commerzbank, KfW Bankengruppe, and DZ Bank. The Deutsche Börsen AG is in Frankfurt. Such clusters support the development of innovations, and this is important, especially in the tertiary sector. There are also advantages because such clusters guarantee a sufficient supply of qualified labour forces, and this is important for companies in that sector.

Another not inconsiderable location factor is the living standard of Frankfurt. The schools and other education institutes are also contributing to the living standard. As stated, the possibilities for education in Frankfurt are very good. The availability and quality of real estate play an important role. Many luxurious apartments and office buildings are located in Frankfurt. There is real estate for every type of worker. Cheaper apartments can be found in the suburbs. In the centre, there is real estate for wealthy people, one of the target groups of the companies, and extremely high-skilled workers. Such properties are modern and luxurious apartments, houses or penthouses. Nevertheless, the city of Frankfurt is still determined to improve the situation. The city is working on 32 urban development projects[31]. These projects go from houses in the rural part to high-rise residential buildings and offices in the

---

[29] Cf. Planungsverband Ballungsraum Frankfurt/Rhein-Main, Wissensatlas Frankfurt, Frankfurt am Main, 2009, page 37-40
[30] Cf. Kirk, 2009, page 11
[31] Cf: http://www.stadtplanungsamt-frankfurt.de/projekte_4092.html, status as of February 25, 2018

centre, which should complement the skyline. This steady development is important to create long-lasting availability of high-quality real estate. Jobs from economic sciences, like investment bankers, for example, are demanded in Frankfurt. This target group has high demands to fulfil in Frankfurt to attract them.

The housing quality is not the only important soft location factor. The recreational and cultural value is important as well because they attract people to settle down in Frankfurt. About 52% of the city's area is a green or leisure area. More than 40 parks and 50 seas and ponds are located in Frankfurt.[32] These locations can be used for sports and recreation. Even though many parks offer room for sports, there are, in addition, 420 sports clubs with 204.163 members[33].

5. Economic structure

5.1. Sectoral structure and economic statistics

To answer the final part of the central question, it is essential to analyse the economic structure and the key branches with their previous developments to recognize a trend. The knowledge acquired from that analysis may help to answer the question if the potential of Frankfurt is exploited. Beforehand, the individual sectors should be defined succinct to avoid misunderstandings in the following text. The primary sector covers agriculture, forestry, and fishery. The secondary sector includes the production industry, and the term tertiary sector contains the service sector.

There were 471.168 employees in 1989[34] in Frankfurt. As shown in figure 8, the service sector was the focus in 1989, with about 71,08%. Nevertheless, the share of the production industry, which was 134.352 or about 28,73% of the labour force, should not be disregarded. The share of the employees in the primary sector, agriculture, forestry and fishery, was really small. 914 people were working in that sector. This was about 0,19% of the labour force.

Figure nine shows the sectoral structure of the year 2015 in Frankfurt and in the whole economic region Frankfurt-Rhein-Main. The inner ring represents the

---

[32] Cf: https://frankfurt-greencity.de/umwelt-frankfurt/frankfurt-die-gruene-stadt/, status as of February 20, 2018

[33] Cf. Stadt Frankfurt am Main, Bürgeramt, Statistik und Wahlen. Statistisches Portrait / Statistical portrait Frankfurt am Main 2016, Frankfurt am Main, 2017, page 5-6

[34] Stadt Frankfurt am Main, Bürgeramt, Statistik und Wahlen. Beschäftigungsschwerpunkte in Frankfurt am Main – das statistisch erfassbare Bild wird vollständiger , Frankfurt am Main, 2015, page 74

sectoral structure of Frankfurt am Main, and the outer ring represents the sectoral structure of the economic area Frankfurt-Rhein-Main.[35]

Figure ten shows the sectoral structure of Frankfurt-Rhein-Main in 2012. It is recognizable that the share of the services is still the largest share of the sectoral structure in Frankfurt. It was 71,08% in 1989. Compared to the statistic of 1989, the share of the tertiary sector grew by 17,92%. The secondary sector decreased by 17,73% from 1989 to 2015 in Frankfurt. The structure of the whole economic area is similar. The tertiary sector is the sector with the largest share. But the share is not as large as the share in Frankfurt am Main. This shows the tertiary sector is more important in Frankfurt than in the whole economic area. This result was predictable because the area of Frankfurt-Rhein-Main is larger than the area of Frankfurt am Main, and there are not such clusters everywhere, such as those in Frankfurt, in the economic area. The share of the primary sector is also small in 2012 and 2015. The share of the primary sector in Frankfurt-Rhein-Main is about 0,3% and is even smaller in Frankfurt am Main with only approximately 0,1%. These developments illustrate the workers move from declining sectors to the successful ones. The structures of 2012 and 2015 are almost equal because there were new workers in total, and the relative amount could remain stable. There is only a small discrepancy in the values of the secondary and tertiary sector. The tertiary sector grew by 0,4%. That shows the economy was stable in recent years, and the sectors are still slowly shifting.

The gross domestic product per capita of Frankfurt-Rhein-Main was 75.816€ in 2012 and increased to 79.313€ in 2014. The gross domestic product per capita of Frankfurt am Main increased from 91.983€ in 2012 to 98.042€ in 2014. This is high, especially the value of Frankfurt am Main, considering the gross domestic product per capita of Germany was 36,064€[36]. The economic growth of 2013 and 2014 can be calculated with those values and can be seen in figure twelve. The economic growth rate increases from 1,88% in Frankfurt-Rhein-Main and from 2,93% in Frankfurt am Main in 2013 to 2,68% in Frankfurt-Rhein-Main and to

---

[35] Source of data: Stadt Frankfurt am Main, Bürgeramt, Statistik und Wahlen, Strukturdatenatlas FrankfurtRheinMain, Frankfurt am Main, 2015 and Stadt Frankfurt am Main, Bürgeramt, Statistik und Wahlen, Strukturdatenatlas Frankfurt, Frankfurt am Main, 2015
[36] Cf: https://www.focus-economics.com/country-indicator/germany/gdp-per-capita-EUR, status as of February 23, 2018

3,55% in Frankfurt am Main. These values underline the solid economic performance and development.

The average gross salary was 35.412€ in Frankfurt-Rhein-Main and 42.785€[37] in Frankfurt am Main in 2014. As a comparison, the average gross salary in Germany was 31.600€[38] in 2014. This again confirms the economic success of the economic area Frankfurt-Rhein-Main.

5.2. Major branches of Frankfurt am Main

The economy of Frankfurt is crucial for the economy. The gross domestic product of the respective city in Frankfurt-Rhein-Main can be calculated with the gross domestic product per capita and the labour force of the respective city. Figure 13 shows the results. Frankfurt had the greatest share of the total gross domestic product with 29,81% in 2014. Therefore, it is suitable to examine major branches of Frankfurt am Main to make an image of the important branches.

Before going into detail of the certain branches, it is interesting to focus on some general facts about the employment in Frankfurt. The most common type of employment is the full-time employment. Almost 60% of the labour force is working in such employments. 25,5% of the workers earn a salary higher than 5.500€.

The financial services are the most important branch in Frankfurt. There are almost no side jobs or mini-jobs and few young workers. This shows the experience is an important requirement and very important in Frankfurt. The intensity and requirements must be high because of the large full-time employment.

The workers are mostly highly educated or have learned their knowledge from an apprenticeship. 14,8% of the workers are classified as experts. The median of the salary is 5663€. More than the half, 55,4%, of the employees, earn more than 5.500€ and very few people, 1,6%, earn less than 2.063€. This shows the highly educated employees must earn high wages because more than half earns more than 5500€ and the median of 5630€. The salary of this branch is the highest in Frankfurt. The most popular professions are insurance and financial services.

The next important branch is the company management and consultancy. Most employers are full-time employed and commuters. Most employees are graduated and

---

[37] Cf: Stadt Frankfurt am Main, Bürgeramt, Statistik und Wahlen, Strukturdatenatlas FrankfurtRheinMain, Frankfurt am Main, 2015 and Stadt Frankfurt am Main, Bürgeramt, Statistik und Wahlen, Strukturdatenatlas Frankfurt, Frankfurt am Main, 2015
[38] Cf: https://de.statista.com/statistik/daten/studie/164047/umfrage/jahresarbeitslohn-in-deutschland-seit-1960/, status as of February 24, 2018

underline the high requirements of this branch. More than a third of the employees can be classified as experts. This is reflected by the high salary. 40,3% earn more than 5.500€ and only 5% earn less than 2.063€. The median of the salary is 5.007€, which is very high, although it is not as high as the one of the finance services. The most popular professions are company consulting and organisation.

The next important branch this text deals with is the services related to information technology. Full-time employed employees, 88%, is in no other branch as high as in the information technology branch. The number of commuters, 73,6%, is also high and higher than the amount of the commuters in the financial services or consultancy. The majority of the employees in this branch completed higher education and are well-educated. There are many specialists and even more experts. This branch has the highest requirements. All these reasons justify the high salaries in this branch. 45,2% of the employees earn more than 5.500€, whereas only 2,5% of the employees earn 2.063€ or less. This is the branch with the second highest number of employees with a salary higher than 5.500€. Only the finance branch pays more to their employees. Popular professions are IT Systems Analysis, application consulting and IT sales. Some employees of that branch even work as a company consultant, which is the main profession in the branch of company management and consultancy. This shows this branch is universal.[39]

6. Conclusion

Frankfurt am Main is the city in Frankfurt-Rhein-Main with the highest importance. Almost one third of the gross domestic product is generated in Frankfurt am Main. It is not the only city contributing to economic performance. Other cities are also contributing more than five billion Euro to the total gross domestic product. Nevertheless, the economic area still depends on Frankfurt because of the high share of the economic performance. After the analysis of the location factors and the economic structure, we can answer the first part of the central question of this paper. To sum up, we can say Frankfurt fulfils many important location factors that companies of the tertiary sectors are looking for. This includes a central location, a good infrastructure, an excellent digital infrastructure, great reputation, education,

---

[39] Cf: Stadt Frankfurt am Main, Bürgeramt, Statistik und Wahlen. Beschäftigungsschwerpunkte in Frankfurt am Main – das statistisch erfassbare Bild wird vollständiger , Frankfurt am Main, 2015, page 75, 79, 83

continuous graduations in demanded subjects, agglomeration benefits, and top real estate. All these location factors attract many companies and workers and definitely contribute to the successful development of an economic. These factors are the main reason companies settle down in Frankfurt. One definitely cannot doubt the success of the branches in Frankfurt, which contribute to great extents to the economic performance of Frankfurt am Main and thus to the economic performance of Frankfurt-Rhein-Main.

But is the region going to succeed and is the potential exploited? First, this question cannot be answered for sure. But there is a positive trend. The city Frankfurt has developed to a valuable location with a high reputation. The gross domestic product per capita has been decreasing for years. There is no sign it may stop increase or decrease. But it is important not to rest on this success and the good strategic location. The economic area must stay attractive to companies. There must be constant improvements of the living space to cope with the rising number of inhabitants. Frankfurt made a good step with the several urban development programs. Investments in other areas may be helpful. The soft location factors must remain attractive, otherwise the valuable employees may relocate. This would be a problem for the area because the tertiary sector is the focus of the economy. For that sector, the quality of the labour is important. There would not be innovations without well-educated employees, and without new innovations or progress, there cannot be sustainable success.

It must be noted that the mentioned location factors are not the only ones in Frankfurt. Every city and district could be analysed in great detail to give us a more precise result, but this would not be realisable in this research paper because of the page limit.

7. Bibliography

Literature:

Palme, G. & Musil, R.: Wirtschaftsgeographie, Braunschweig 2012[1]

Kirk, C.: Wirtschaftsstandort Frankfurt am Main, Darmstadt 2001

Kirk, C.: Wirtschaftsstandort Frankfurt am Main, Darmstadt 2009

Stadt Frankfurt am Main, Bürgeramt, Statistik und Wahlen: Statistisches Portrait / Statistical portrait Frankfurt am Main 2016, Frankfurt am Main, 2017
Accessible via https://www.frankfurt.de/sixcms/media.php/678/JB2017_Statistisches%20Portrait%202016.pdf

Industrie- und Handelskammer: IHK-BEZIRK FRANKFURT IN ZAHLEN 2016|2017, Frankfurt am Main, 2017
Accessible via https://www.frankfurt-main.ihk.de/images/broschueren/IHK-Bezirk%20in%20Frankfurt%20Zahlen%202016_2017.pdf

IHK-Forum Rhein-Main: FrankfurtRheinMain in Zahlen 2018, Hanau, 2018
Accessible via https://www.frankfurt-main.ihk.de/images/broschueren/FRM_zahlen_2018(D).pdf

Stadt Frankfurt am Main, Bürgeramt, Statistik und Wahlen: Klassifikation der Berufe 2010: Erste Ergebnisse zu den Beschäftigten in Frankfurt am Main am 30. Juni 2013 (Teil 2), Frankfurt am Main, 2014
Accessible via https://www.frankfurt.de/sixcms/media.php/678/02_Beschäftigte_Berufsklassifikation_Teil2.pdf

Stadt Frankfurt am Main, Bürgeramt, Statistik und Wahlen. Beschäftigungsschwerpunkte in Frankfurt am Main – das statistisch erfassbare Bild wird vollständiger , Frankfurt am Main, 2015
Accessible via https://www.frankfurt.de/sixcms/media.php/678/2015-4_Beschäftigungsschwerpunkte.pdf

Planungsverband Ballungsraum Frankfurt/Rhein-Main, Wissensatlas Frankfurt, Frankfurt am Main, 2009

Accessible via https://www.frankfurt.de/sixcms/media.php/738/Wissensatlas-FRM2009.pdf

Websites:

https://www.region-frankfurt.de/Regionalverband/Region-in-Zahlen/Statistik-Viewer-Regionalverband/in-dex.php?mNavID=2629.36&sNavID=2629.36&La=1#a2, status as of February 25, 2018

http://www.frankfurt.de/sixcms/detail.php?id=2556771, status as of February 25, 2018

http://frankfurt-business.net/standort-frankfurt/infrastruktur, status as of February 25, 2018

https://exporo.de/standortanalyse/frankfurt-am-main, status as of February 25, 2018

http://global-internet-map-2012.telegeography.com, status as of February 25, 2018

https://data.worldbank.org/indicator/NY.GDP.PCAP.CD?end=2016&loca-tions=DE&start=1970&view=chart, status as of February 25, 2018

http://frankfurt-business.net/standort-frankfurt/zahlen-daten-fakten/bildung, status as of February 25, 2018

http://www.stadtplanungsamt-frankfurt.de/projekte_4092.html, status as of February 25, 2018

https://frankfurt-greencity.de/umwelt-frankfurt/frankfurt-die-gruene-stadt, status as of February 25, 2018 https://www.focus-economics.com/country-indicator/germany/gdp-per-capita-EUR, status as of February 25, 2018

https://de.statista.com/statistik/daten/studie/164047/umfrage/jahresarbeitslohn-in-deutschland-seit-1960/, status as of February 25, 2018

other sources:
online service "Google Maps" by Google, accessible via http://maps.google.com/

Stadt Frankfurt am Main, Bürgeramt, Statistik und Wahlen: Strukturdatenatlas Frankfurt, Frankfurt am Main, 2015
Accessible via http://www.statistik.stadt-frankfurt.de/strukturdatenatlas/stadtteile/html/atlas.html

Online service "Fachkräftemonitor Hessen" by the Chamber of Commerce and Industry, accessible via http://www.fachkraefte-hessen.de/fachkraeftemonitor.html#ivmhirej2vik

Stadt Frankfurt am Main, Bürgeramt, Statistik und Wahlen, Strukturdatenatlas FrankfurtRheinMain, Frankfurt am Main, 2015 and Stadt Frankfurt am Main, Bürgeramt, Statistik und Wahlen, Strukturdatenatlas Frankfurt, Frankfurt am Main, 2015
Accessible via http://www.statistik.stadt-frankfurt.de/strukturdatenatlas/region_rheinmain/html/atlas.html

8. Annex

List of figures:

Cover photo:

Extracted: http://img.fotocommunity.com/frankfurt-skyline-bei-nacht-neu-e7e322a7-db6e-4bf7-a3ad-4c204922cbca.jpg?width=1000, status as of February 21, 2018

1:

extracted: http://www.firmendb.de/info-pool/metropolregionen.php, status as of February 13, 2018

2:

Source of data: Stadt Frankfurt am Main, Bürgeramt, Statistik und Wahlen, Strukturdatenatlas Frankfurt, Frankfurt am Main, 2015

Independent cities	Federal state
Frankfurt am Main	Hesse
Offenbach am Main	Hesse
Wiesbaden	Hesse
Mainz	Rhineland-Palatine
Worms	Rhineland-Palatine

Darmstadt	Hesse
Aschaffenburg	Bavaria
Districts:	
Main-Taunus district	Hesse
Hochtaunus district	Hesse
Wetterau district	Hesse
Main-Kinzig district	Hesse
Offenbach district	Hesse
Groß-Gerau district	Hesse
Aschaffenburg district	Bavaria
Miltenberg district	Bavaria
Darmstadt-Dieburg district	Hesse
Odenwald district	Hesse
Bergstraße district	Hesse
Alzey-Worms district	Rhineland-Palatine
Mainz-Bingen district	Rhineland-Palatine
Rheingau-Taunus district	Hesse
Limburg-Weilburg district	Hesse
Gießen district	Hesse
Vogelsberg district	Hesse
Fulda district	Hesse

3:

Extracted: https://www.frankfurt-main.ihk.de/images/broschueren/The%20Frankfurt%20CCI%20District%20in%20figures%202016_2017.pdf, status as of February 18, 2018

4:

Extracted: http://global-internet-map-2012.telegeography.com, status as of February 18, 2018

5:

Source of data: "Fachkräftemonitor Hessen" provided by the Chamber of Commerce and Industry, 2017

Description:

Lack of skilled labour, presented in a bar chart, sorted according different occupational groups

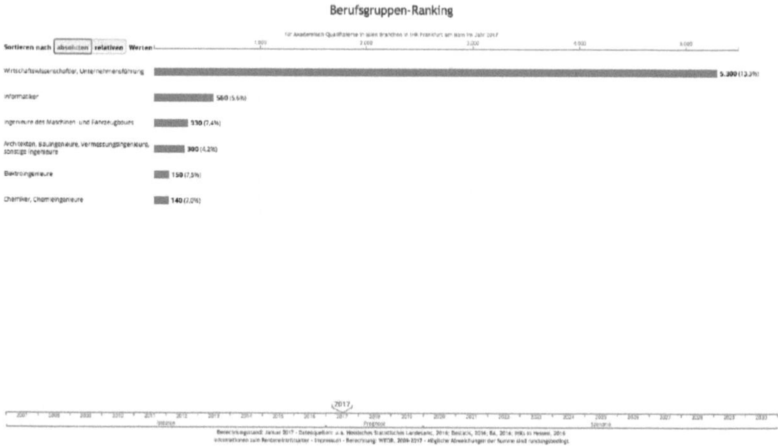

6:

Source of data: "Fachkräftemonitor Hessen" provided by the Chamber of Commerce and Industry, 2017

Description:

Supply and demand for economic sciences and management, presented in a graph

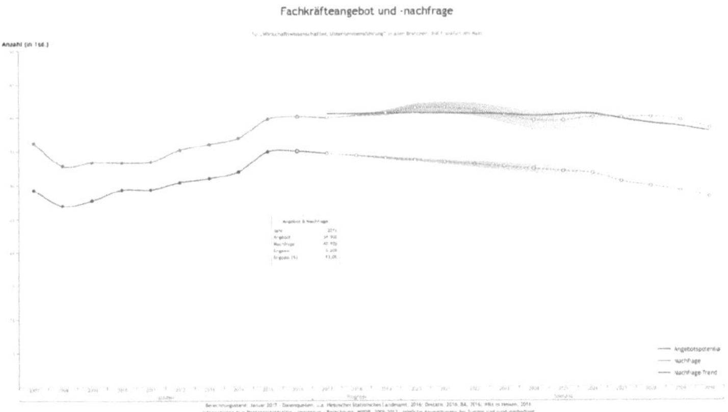

7:

Source of data: "Fachkräftemonitor Hessen" provided by the Chamber of Commerce and Industry, 2017

Description:

Supply and demand for jobs in information technology, presented in a graph

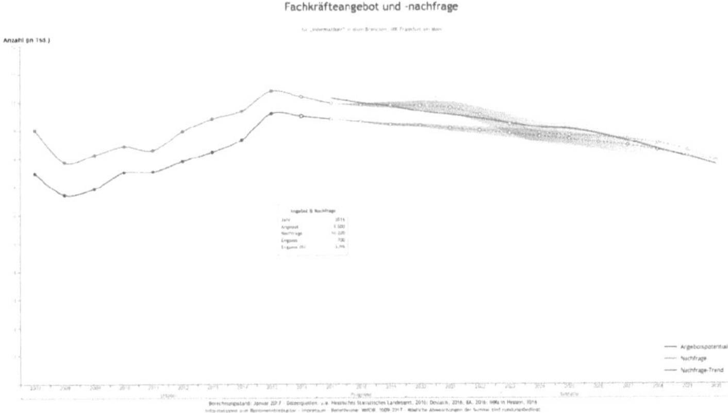

8:

Source: Stadt Frankfurt am Main, Bürgeramt, Statistik und Wahlen. Beschäftigungsschwerpunkte in Frankfurt am Main – das statistisch erfassbare Bild wird vollständiger , Frankfurt am Main, 2015, page 74

Description:

Employees contributing to social security in mid-1989 by economic sectors

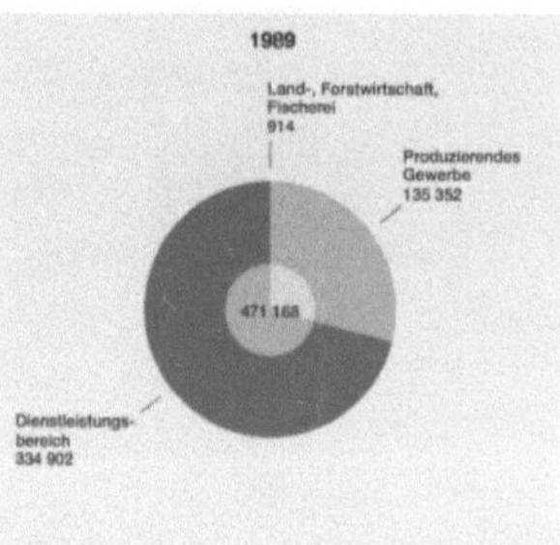

9:

Created in accordance with data

Source of data: Stadt Frankfurt am Main, Bürgeramt, Statistik und Wahlen, Strukturdatenatlas FrankfurtRheinMain, Frankfurt am Main, 2015 and Stadt Frankfurt am Main, Bürgeramt, Statistik und Wahlen, Strukturdatenatlas Frankfurt, Frankfurt am Main, 2015

Description:

Sectoral structure of Frankfurt am Main and Frankfurt-Rhein-Main from in 2015

Inner ring: Frankfurt am Main; outer ring: Frankfurt-Rhein-Main

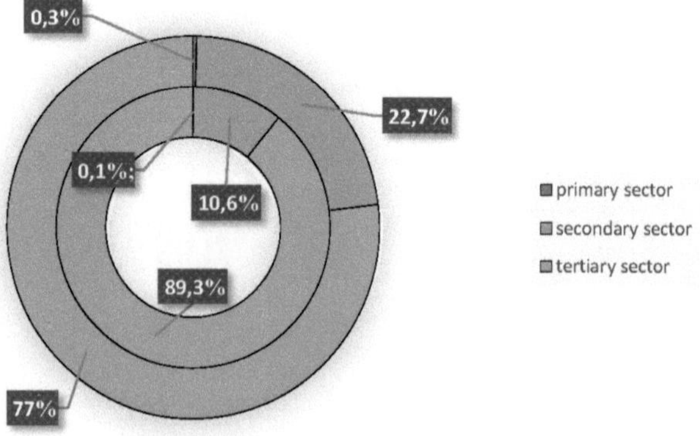

10:

Created in accordance with data

Source of data: Stadt Frankfurt am Main, Bürgeramt, Statistik und Wahlen, Strukturdatenatlas FrankfurtRheinMain, Frankfurt am Main

Description:

Sectoral structure of Frankfurt-Rhein-Main from in 2012

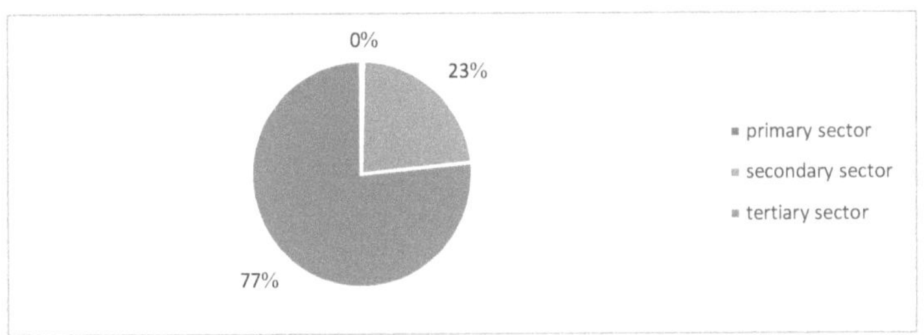

11:

Created on my own in accordance with data

Source of data: Stadt Frankfurt am Main, Bürgeramt, Statistik und Wahlen, Strukturdatenatlas FrankfurtRheinMain, Frankfurt am Main, 2015 and Stadt Frankfurt
am Main, Bürgeramt, Statistik und Wahlen, Strukturdatenatlas Frankfurt, Frankfurt
am Main, 2015

Description:

Gross domestic product per capita in Frankfurt am Main and Frankfurt-Rhein-Main
from 2012 to 2014

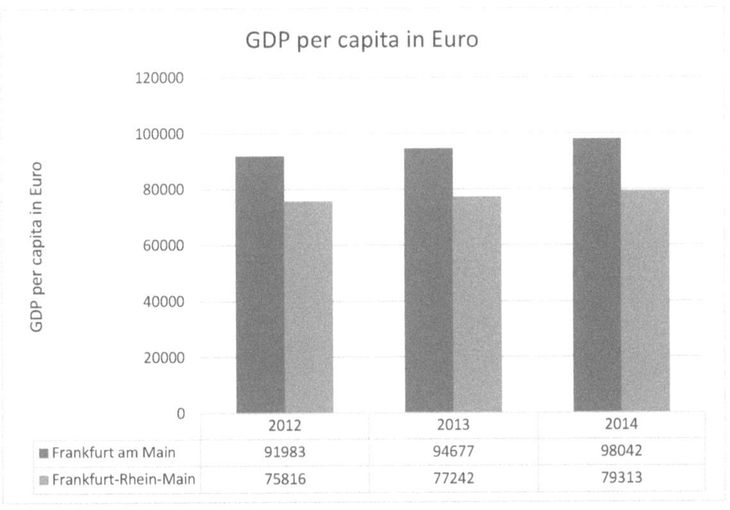

12:

Calculated on my own in accordance with data

Source of data: Stadt Frankfurt am Main, Bürgeramt, Statistik und Wahlen, Strukturdatenatlas FrankfurtRheinMain, Frankfurt am Main, 2015 and Stadt Frankfurt am Main, Bürgeramt, Statistik und Wahlen, Strukturdatenatlas Frankfurt, Frankfurt am Main, 2015

Description:

Economic growth rate of Frankfurt am Main and Frankfurt-Rhein-Main in 2013 and 2014

Economic growth rate	2013	2014
Frankfurt am Main	2,99%	3,55%
Frankfurt-Rhein-Main	1,88%	2,68%

13:

Calculated on my own in accordance with data

Source of data: Stadt Frankfurt am Main, Bürgeramt, Statistik und Wahlen, Strukturdatenatlas FrankfurtRheinMain, Frankfurt am Main, 2015 and Stadt Frankfurt am Main, Bürgeramt, Statistik und Wahlen, Strukturdatenatlas Frankfurt, Frankfurt am Main, 2015

Description:

Total gross domestic products and the associated share of the total gross product for the respective member cities and districts of Frankfurt-Rhein-Main

city / district	gross domestic product in Euro	share of total gross domestic product in %
Aschaffenburg	3.389.752.551,00	1,92
Darmstadt	7.546.786.491,00	4,27
Frankfurt am Main	52.720.222.702,00	29,81
Mainz	7.971.413.940,00	4,51
Offenbach am Main	3.093.986.949,00	1,75
Wiesbaden	11.283.530.434,00	6,38
Worms	2.049.833.181,00	1,16
Alzey-Worms district	1.862.013.328,00	1,05
Aschaffenburg district	3.415.562.376,00	1,93
Bergstraße district	4.972.280.975,00	2,81
Darmstadt-Dieburg distr	4.768.427.360,00	2,70
Fulda district	5.192.818.064,00	2,94
Gießen district	5.550.744.200,00	3,14
Groß-Gerau district	8.188.177.942,00	4,63
Hochtaunus district	7.310.153.100,00	4,13
Limburg-Weilburg distric	2.922.056.505,00	1,65
Main-Kinzig district	8.726.911.336,00	4,93
Main-Taunus district	8.814.503.298,00	4,98
Mainz-Bingen district	4.136.405.146,00	2,34
Miltenberg district	2.623.055.240,00	1,48
Odenwald district	1.588.240.381,00	0,90
Offenbach district	9.029.085.575,00	5,10
Rheingau-Taunus distric	2.687.623.393,00	1,52
Vogelsberg district	1.681.491.360,00	0,95
Wetterau district	5.344.895.365,00	3,02
total	176.869.971.192,00	100,00

YOUR KNOWLEDGE HAS VALUE

- We will publish your bachelor's and
  master's thesis, essays and papers

- Your own eBook and book -
  sold worldwide in all relevant shops

- Earn money with each sale

Upload your text at www.GRIN.com
and publish for free